P9-ASC-205

CORDUROY'S CHRISTMAS SURPRISE

Grosset & Dunlap

A bear's share of the royalties from the sale of *Corduroy's Christmas Surprise* goes to the Don and Lydia Freeman Research Fund to support psychological care and research concerning children with life-threatening illness.

Some material in this book was first published in *Corduroy's Christmas* in 1992 by Viking, a division of Penguin Putnam Inc.

 Published by Grosset & Dunlap, a division of Penguin Putnam Books for Young Readers, New York. GROSSET & DUNLAP is a trademark of Penguin Putnam Inc. Published simultaneously in Canada. Printed in the U.S.A. Illustrations created by Lisa McCue.

Library of Congress Cataloging-in-Publication Data

Corduroy's Christmas surprise / based on the character created by Don Freeman; illustrated by Lisa McCue.
p. cm. — (Reading railroad books)
Summary: Corduroy's letter to Santa is full of requests for his friends, who will be spending Christmas Day with him, but Santa has a surprise in store for a very special bear, too.
[1. Christmas—Fiction. 2. Gifts—Fiction. 3. Teddy bears—Fiction. 4. Toys—Fiction.] I. Freeman, Don. II. McCue, Lisa, ill. III. Series.

PZ7.C815358 2000
[E]—dc21 00-063642

ISBN 0-448-42191-7 D E F G H I J

Corduroy's Christmas Surprise

Based on the character created by DON FREEMAN
Illustrated by LISA McCUE

Grosset & Dunlap, Publishers

There was only a week to go until Christmas, and Corduroy could hardly wait! This was one of the best times of the year. Corduroy loved the smell of his Christmas tree. He loved taking the ornaments and lights out of the closet.

He loved stringing popcorn and berries and trimming the tree.

And that was only the beginning. Corduroy's list of fun things to do at Christmastime went on and on and on.

SATIN

This year, Corduroy thought to himself, Christmas was going to be even more fun than usual! His best friends Mouse, Rabbit, Dolly, and Puppy were coming over to spend Christmas Day at Corduroy's house. And he had little gifts for his guests, so he pulled out the wrapping paper, tape, and ribbons. Wrapping presents always got Corduroy into the holiday spirit.

Dear Santa,
Please bring me
Ice Skates
Sweater
Trains
Ball
Thank you
CORDUROY

That reminded Corduroy that he hadn't written his letter to Santa! What did he want for Christmas? Well, thought Corduroy, it would be fun to have ice skates this winter. He could use a new warm sweater. Oh! He had always wanted a train set! And a new ball would be nice, too! Corduroy wrote his letter in his neatest, best printing.

Just then, the doorbell rang—*ding-dong!* Corduroy put down his letter and went to the door, and there were Mouse, Rabbit, and Puppy standing in front of his house. And they were singing! *We wish you a Merry Christmas and a Happy New Year!*

When the carolers had finished, Puppy said, "Corduroy, grab your coat and come with us! We'll go caroling at Dolly's house next!"

So Corduroy did just that. He put on his coat and scarf, and the four friends made their way over to Dolly's house. As they began to sing, Dolly opened the door. She looked so surprised!

Then Dolly invited them all inside. "I'm putting the finishing touches on my gingerbread house," she said. "Why don't you four come in and help me?"

A real homemade gingerbread house?! Corduroy had to see this!

And what a gingerbread house it was! It had lots of windows, a fancy front door, *and* a chimney. Corduroy could hardly believe his eyes!

"This is one of the best things about Christmas," said Dolly.

"Don't forget Christmas trees!" said Corduroy.

"And Christmas carols!" said Rabbit.

"And Christmas candy!" said Puppy, as he gobbled some candy.

"And Christmas presents!" said Mouse.

Everybody laughed. Then they started talking about what each of them wanted for Christmas. Dolly said she wished for a pair of ice skates. Puppy said he'd always wanted a train set.

Rabbit said he could use a new sweater. And Mouse said a brand-new ball would be fun.

Corduroy didn't say anything. How could he tell them that he had written to Santa to ask for *all* of those things?

Back at home, Corduroy found his letter to Santa right where he had left it. He decided to make a few changes.

What if there weren't enough presents to go around? What if Corduroy got what he wanted, but his friends didn't?

So Corduroy pulled out a new piece of paper and wrote another letter to Santa. This is what it said:

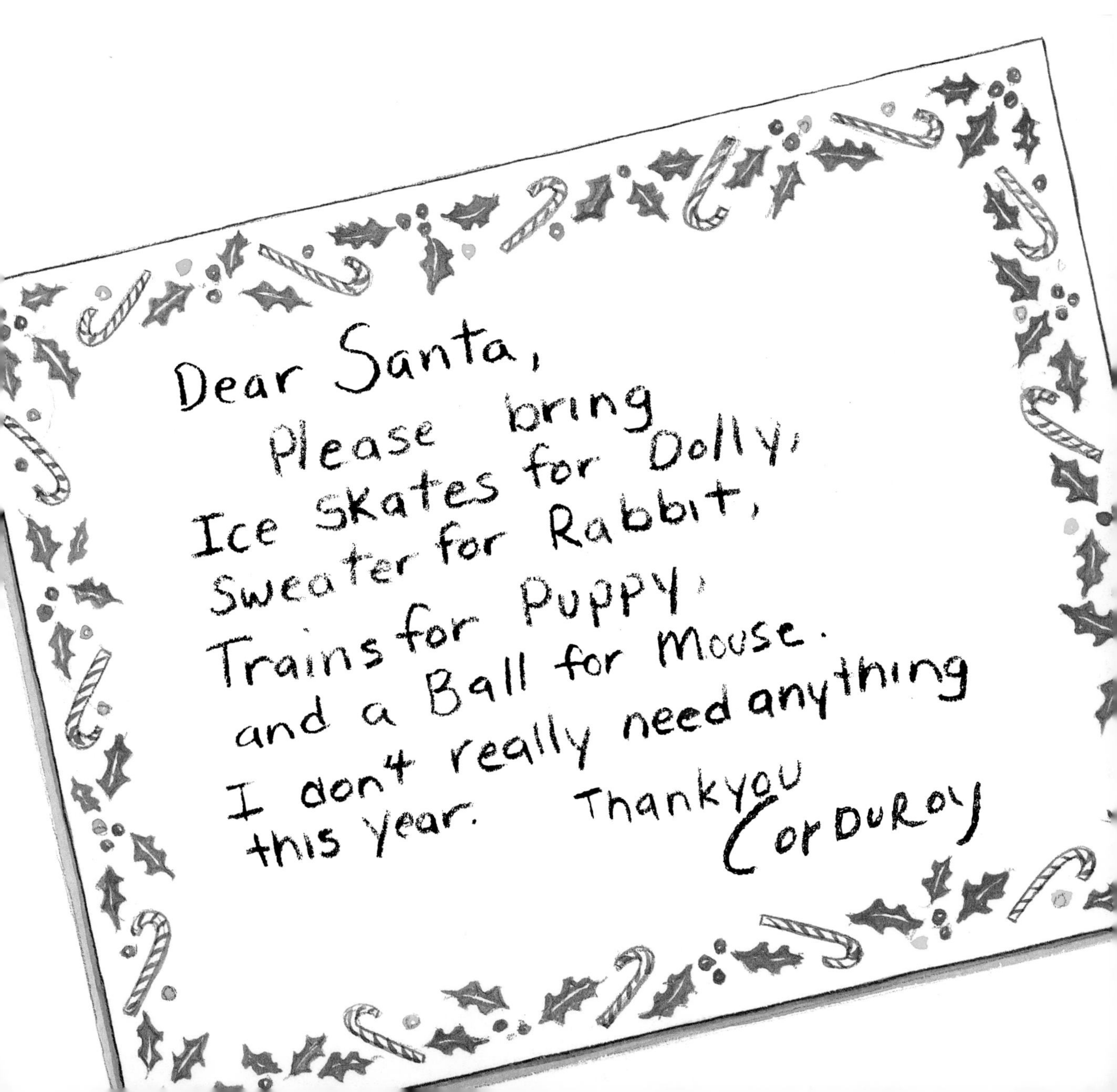

Dear Santa,
Please bring
Ice Skates for Dolly,
Sweater for Rabbit,
Trains for Puppy,
and a Ball for Mouse.
I don't really need anything
this year. Thankyou
Corduroy

Corduroy put the letter in an envelope. In his neatest, best printing, he addressed it to Santa Claus at the North Pole. He stuck on a stamp and sealed the letter shut.

Then Corduroy pulled on his coat, his scarf, and his boots…

...and he walked outside to the mailbox and dropped the letter inside. He hoped Santa would get it before Christmas!

A few days later, it was Christmas Eve! The very next morning, Corduroy's friends would be coming over to celebrate with him. Corduroy was getting ready by baking Christmas cookies. He hoped he had made enough!

TEA
SUGAR
FLO
COOKIES
LISA MCCUE
1992

Before Corduroy knew it, it was time for bed. But first, he had two important things to do. He hung his stocking on the mantel…

...and he left milk and cookies on the table for you-know-who. Maybe Santa wouldn't be bringing anything for him, Corduroy thought. But he was so happy that his friends would get exactly what they wanted!

Then Corduroy climbed into bed. All night long, he dreamed of the happy looks on his friends' faces when they opened their presents on Christmas morning.

For Santa
Merry
Christmas

The next morning, Corduroy sat up in bed and rubbed his eyes. It was Christmas Day! It had finally come! Corduroy looked out the window to see a crisp, sunny day and a fresh blanket of snow on the ground.

TO DOLLY
TO PUPPY
TO MOUSE
TO RABBIT

Then Corduroy rushed out to the living room to find lots of presents under the tree! Corduroy could see there was one labeled for Mouse, and one for Rabbit. There was one for Dolly, and one for Puppy. And—could it be?—there was one for Corduroy, too, and extra presents for each of them! Corduroy was so surprised. Now he *really* couldn't wait for his friends to arrive.

Christmas

Once everyone had come, Mouse, Rabbit, Dolly, Puppy, and Corduroy sat down together by the tree and opened their presents. Dolly got ice skates and a scarf. Rabbit got the sweater he wanted, and a picture book, too! Puppy got the train set of his dreams and a new hat. Mouse got a ball and some warm winter boots.

And Corduroy got ice skates and a sweater! How in the world did Santa know that he wanted them?

But really, Corduroy thought to himself, the best Christmas gift of all was that he was surrounded by his friends on Christmas Day—and they were all so happy!

That made Corduroy happy, too. He would have to write one more letter to Santa—to say thank you for the best Christmas ever.